The Quantum Breakthrough

How Quantum Computing is Transforming Technology

Emma Royce Smartley

Copyright © 2024 Emma Royce Smartley

All rights reserved.

DEDICATION

To those who have the audacity to envision a world reshaped by invention and exploration.

The visionaries, researchers, and trailblazers in the field of quantum computing, whose unwavering quest for knowledge keeps expanding the realm of what is conceivable, are honored in this book. We are all motivated to strive for the remarkable by your curiosity and dedication.

I want to express my gratitude to my family and friends for their continuous encouragement and support throughout this journey. Your admiration for my work has continuously inspired me.

And to the readers, I hope that my investigation into quantum computing will spark your creativity and encourage you to embrace technology's boundless possibilities. Let's work together to imagine a more promising future influenced by scientific advancements and discoveries.

CONTENTS

ACKNOWLEDGMENTS...1

CHAPTER 1..1

An Overview of Quantum Computation......................................1

   1.1 Comprehending Quantum vs Classical Computing..........1

   1.2 Quantum Mechanics Fundamentals....................................4

   1.3 A Synopsis of Quantum Computing's Development and History.....6

CHAPTER 2..9

Quantum Gates and Quantum Bits (Qubits)................................9

   2.1 Qubits: The Foundational Elements...................................9

   2.2 Qubit Superposition and Entanglement............................12

   2.3 Quantum Circuits and Gates..14

CHAPTER 3..18

Algorithms in Quantum..18

   3.1 The Algorithm of Shore..18

   3.2 The Algorithm of Governor..20

   3.3 The Potential of Other Quantum Algorithms...................22

CHAPTER 4..26

Security and Quantum Cryptography.......................................26

   4.1 Classical Cryptography and Quantum Dangers...............26

   4.2 Post-Quantum Cryptography...29

   4.3 Quantum Key Distribution (QKD).....................................31

CHAPTER 5..36

Drug Discovery Using Quantum Computing............................36

5.1 Molecular Structure Simulation..36

5.2 Quickening Drug Development and Evaluation............................39

5.3 Biotechnology and Genomics: Quantum Applications..................41

CHAPTER 6..**45**

Optimization Problems with Quantum Computing......................**45**

6.1 Financial Optimization..45

6.2 Logistics and Supply Chain Optimization.....................................48

6.3 Optimization via Quantum Machine Learning..............................51

CHAPTER 7..**54**

Physical Implementation and Quantum Hardware......................**54**

7.1 Qubits that are superconducting..54

7.2 Other Qubit Technologies and Trapped Ions.................................57

7.3 Quantum Hardware Challenges...59

CHAPTER 8..**63**

Programming Languages and Quantum Software.......................**63**

8.1 Frameworks for Quantum Programming.......................................63

8.2 Ecosystem of Quantum Software..67

8.3 The Development of Quantum Algorithms...................................69

CHAPTER 9..**73**

Quantum Computing Applications in Industry............................**73**

9.1 Medical Applications of Quantum Computing..............................73

9.2 Energy Quantum Computing...76

9.3 Artificial Intelligence and Quantum Computing...........................79

CHAPTER 10..**83**

Quantum Computing's Future...**83**

10.1 Recent Developments and Upcoming Milestones.........................83

10.2 Future Prospects: Quantum Domination and Beyond..................87

10.3 Implications for Society and Ethics...89

ABOUT THE AUTHOR..93

ACKNOWLEDGMENTS

To everyone who helped create **The Quantum Breakthrough**, I would like to extend my sincere gratitude. Without many people's help, encouragement, and knowledge, this book would not have been possible.

I want to start by sincerely thanking my mentors and colleagues in the quantum computing sector. My comprehension and enthusiasm for this intricate topic have been greatly influenced by your advice and thoughts. We are especially grateful to everyone who shared their research and experiences, which helped to make this book better.

My family and friends have been my rock during this trip, and I would also like to thank them for their contributions. Your support and faith in my vision have inspired me to overcome obstacles and stay committed to my objectives.

I am appreciative of the readers who are joining me in exploring the intriguing realm of quantum computing. Your excitement and interest motivate me to present these concepts in an understandable and captivating manner.

Lastly, I want to express my gratitude to the publishing team for their diligence and commitment in making this book a reality. This process has been easy and satisfying because of your professionalism and knowledge.

I am thankful to have each and every one of you by my side as we explore the possibilities of this revolutionary technology. Together, we are on the cusp of an exciting quantum revolution.

CHAPTER 1

AN OVERVIEW OF QUANTUM COMPUTATION

For certain applications, quantum computing promises capabilities that greatly surpass those of traditional computers, marking a paradigm change in the field of computation. This chapter discusses the basic ideas of quantum computing, contrasts it with classical computing, looks at the underlying principles of quantum mechanics, and charts the evolution of this innovative technology across time.

1.1 Comprehending Quantum vs Classical Computing

Fundamentally, the way information is processed is what separates classical and quantum computing.

Traditional Computer Science:

Bits are the smallest unit of data used in classical

computers, and each bit can be in one of two states: 0 or 1. In order to process these bits and conduct computations, binary operations that scale linearly with the amount of data are used in conjunction with traditional logic gates. Even while classical computing has transformed both technology and society, it is limited when dealing with complicated issues, especially those that need complicated computations or enormous datasets.

The concept of quantum computing

In contrast, quantum bits, or qubits, are used in quantum computers. In contrast to traditional bits, qubits have a characteristic known as superposition that allows them to reside in a state of 0, 1, or both at the same time. Quantum computers can process enormous volumes of data at once thanks to these capabilities. For instance:

Two qubits can represent all four states simultaneously, whereas a system of two classical bits can represent the four different states (00, 01, 10, 11).

This characteristic makes it possible for quantum

computers to solve some issues far more quickly than their classical equivalents.

Power Leap in Expansion:

There are two main quantum phenomena responsible for the exponential increase in computing power:

- **Superposition** enables a quantum computer to investigate several solutions at once.

- Another fundamental quantum principle, entanglement, allows entangled qubits to be correlated with each other, meaning that the state of one immediately affects the state of another, regardless of how far away they are. For operations like factorization, database searches, and optimization problems, this can greatly increase computational performance.

Quantum computers have the potential to transform domains where classical approaches fall short, such as materials research, complex system simulation, and cryptography, even though classical computers are superior in a wide range of applications.

1.2 Quantum Mechanics Fundamentals

Several fundamental ideas of quantum physics must be understood in order to comprehend quantum computing:

- **Superposition (-):** The ability of a quantum system to exist in numerous states simultaneously is known as superposition in quantum physics. For example, until it is measured, a qubit may be in a state of 0, 1, or both at the same time. Compared to classical computers, which must process each state sequentially, quantum computers can manage enormous volumes of data more effectively thanks to this feature.

- **Entanglement:** A phenomenon known as entanglement occurs when two or more qubits get intertwined to the point where one qubit's state cannot be explained separately from the states of the others. This implies that regardless of the distance between two qubits, measuring one of them instantly yields information about its entangled partner. Entanglement contributes to the enormous computational capability of quantum systems by improving the ability of quantum algorithms to process information in parallel.

- **Interference:** The mechanism by which quantum states can combine and either cancel or enhance one another is known as quantum interference. This idea is essential to quantum algorithms because it permits some computing methods to increase the likelihood of desired results while decreasing the likelihood of undesirable ones. Quantum algorithms, like Shor's technique for factoring huge numbers, can produce results that are significantly better than those of classical approaches thanks to the utilization of interference.

These ideas collectively serve as the cornerstone of quantum computing. They make it possible for quantum computers to carry out intricate calculations that are impossible with just classical computing, especially in domains like machine learning, cryptography, and optimization.

1.3 A Synopsis of Quantum Computing's Development and History

Beginning in the 1980s, the development of quantum computing has seen numerous noteworthy turning points:

History and Foundational Events:

The scientist Richard Feynman first introduced the idea of quantum computing in 1981 after realizing that quantum systems could not be effectively simulated by classical computers. A theoretical model of a quantum computer was put forth by David Deutsch in 1985, providing a foundation for the possible operation of such a device.

Key Figures:

The first quantum algorithm that could factor huge integers exponentially faster than the most well-known conventional algorithms was created in 1994 by Peter Shor, demonstrating the usefulness of quantum computers for cryptography.

Soon after, Lov Grover presented his search method, which was quadruply faster than traditional algorithms in searching unsorted datasets.

These discoveries sparked additional study and advancement in quantum theory and technology.

Development and Organizations:

Many academic organizations and research centers started funding quantum computing research in the late 20th and early 21st centuries. Prominent companies including Google, IBM, and D-Wave Systems have contributed significantly to the advancement of quantum algorithms and hardware. Google asserted in 2019 that they had achieved quantum supremacy, showing that its quantum processor could outperform the most potent classical supercomputers in a particular calculation.

With developments in quantum hardware, algorithms, and applications, the topic has continued to move quickly, attracting increased interest and funding from the public and private sectors. Looking ahead, quantum computing has the potential to completely alter our conception of computation and its uses in a variety of domains.

CHAPTER 2

QUANTUM GATES AND QUANTUM BITS (QUBITS)

The quantum bit, or qubit, is the basic unit of information in the field of quantum computing. This chapter offers a thorough examination of qubits as the fundamental units of quantum computation, exploring their special characteristics, the physical systems in which they are implemented, the concepts of entanglement and superposition, and the quantum gates that control these qubits in quantum circuits.

2.1 Qubits: The Foundational Elements

The fundamental building blocks of quantum computing are qubits, which are essentially different from classical bits in a number of important ways.

Qubit Definition:

The quantum counterpart of a classical bit is called a qubit.

A qubit can exist in any quantum superposition of 0 and 1, whereas a classical bit can only exist in one of the two states of 0 or 1. Because of this characteristic, quantum computers are able to process information in ways that are not possible for classical computers. A qubit's basis states can be represented mathematically as a linear combination:

$$|\psi\rangle = \alpha|0\rangle + \beta|1\rangle$$

The probability amplitudes of the qubit being in the states $|0\rangle$ or $|1\rangle$ are represented by the complex numbers α and β, respectively. The conditions ($|\alpha|^2 + |\beta|^2 = 1$) must hold to guarantee normalization, and ($|\alpha|^2$) and ($|\beta|^2$) provide the probabilities of measuring the qubit in either state.

The Differences Between Qubits and Classical Bits:

The properties of superposition and entanglement are the primary distinctions between qubits and classical bits. Qubits can encode many states at once, whereas classical bits can only represent one state at a time. Because qubits can do numerous calculations simultaneously, this capacity results in a significant gain in processing power.

Qubit Physical Implementations:

To achieve qubits, a number of physical systems have been created, each with unique benefits and difficulties. Among the most popular implementations are:

- **Superconducting Qubits:** These circuits function at very low temperatures and are composed of superconducting materials. They manufacture qubits using Josephson junctions, in which the direction of current flow determines the quantum state. Because of their ease of integration into larger quantum systems and comparatively quick gate operations, superconducting qubits have gained popularity.

- **Trapped Ion Qubits:** This method uses electromagnetic fields to confine individual ions in a vacuum. Quantum gates are created by manipulating quantum states using lasers. One benefit of trapped ions is their extended coherence times, which enable high-fidelity operations.

- The goal of this method is to employ exotic particles known as anyons, which have braiding statistics.

Topological qubits are a promising approach to fault-tolerant quantum computing since they should be more resilient to errors and noise.

- **Photonic Qubits:** By virtue of their polarization states, photons can be used as qubit representations. Photonic qubits are appropriate for quantum communication and networking applications because they may be sent across great distances with no loss.

These various applications demonstrate the creative methods scientists are using to apply the ideas of quantum physics to real-world computers.

2.2 Qubit Superposition and Entanglement

Understanding the capabilities of qubits and quantum computing requires an understanding of superposition and entanglement.

Explaining Superposition: A key idea in quantum physics, superposition permits qubits to exist in several states at once. In contrast to a classical bit, which can only be in one of these states, a single qubit, for instance, can simultaneously be in both 0 and 1. This characteristic allows exponential computations to be carried out by quantum computers:

- A quantum computer with (n) qubits can represent (2^n) states simultaneously. This is known as parallel processing. A three-qubit system, for example, can concurrently represent eight distinct state combinations (000, 001, 010, 011, 100, 101, 110, and 111). This enables quantum algorithms to investigate several solutions concurrently, resulting in significantly faster computations.

Quantum Entanglement: Entanglement is a special characteristic of quantum systems in which the states of two qubits are connected, independent of their distance from one another. When two qubits are entangled, the state of one is immediately affected by the measurement of the other:

- **Non-Local Correlations:**
  This non-locality goes against traditional notions of

separation and independence. For instance, regardless of their distance from one another, if two qubits are entangled and one is detected to be in the state 0, the other qubit will instantly collapse into the state 1.

- **Quantum Computing Implications:**

    Many quantum protocols and algorithms, such as superdense coding and quantum teleportation, depend on entangled qubits. Richer computational power and more effective information transfer are made possible by the capacity to use entanglement to perform more sophisticated operations than would be feasible with unentangled qubits.

In order to understand how quantum computers can perform better than their classical counterparts in specific computational tasks, it is essential to understand superposition and entanglement.

2.3 Quantum Circuits and Gates

Similar to conventional logic gates but built to operate qubits in accordance with quantum theory, quantum gates are the fundamental components of quantum circuits.

Qubit Manipulation via Quantum Gates:

Qubits are subjected to operations by quantum gates that alter their states. Quantum gates are intrinsically reversible and capable of producing superpositions and entanglement, in contrast to classical gates that carry out deterministic actions. Unitary matrices can be used mathematically to depict quantum gates while maintaining the normalization of the total probability.

Typical Gate Types:

In quantum computation, a number of basic quantum gates are frequently employed, each with a distinct function:

The Hazard Gate (H):

By converting a single qubit from a basic state into an equal superposition of both states, the Hadamard gate produces superposition. For instance, $|0\rangle$ produces the following result when the Hadamard gate is applied:

$$H|0\rangle = \frac{1}{\sqrt{2}}(|0\rangle+|1\rangle)$$

For quantum algorithms that depend on examining several

options, this gate is essential.

The Pauli gates, which are single-qubit gates that rotate around the Bloch sphere's axis, are **Pauli Gates (X, Y, Z):**
- **X Gate:** Flips the qubit's state (i.e., $X|0\rangle=|1\rangle$) and $X|1\rangle=|0\rangle$) in a manner similar to that of a classical NOT gate.
- In addition to the state flip, the Y Gate introduces a phase shift.
- **Z Gate**: Modifies the qubit's phase without affecting its probability by applying a phase flip.

CNOT (Controlled NOT Gate):

When the control qubit is in state 1, the two-qubit CNOT gate flips the state of a target qubit. It is frequently employed in quantum algorithms and is necessary for generating entanglement. The expression for the CNOT operation is as follows:
- If the control qubit is $|0\rangle$, the target qubit stays the same.
- The target qubit reverses its state if the control qubit is $|1\rangle$.

Part in Computations: Complex algorithms can be carried out by combining quantum gates to create quantum circuits. Quantum computers can execute complex tasks with amazing efficiency because the order in which gates are applied to a group of qubits affects the total computation.

Quantum computers use quantum gates and circuits to do computations. Quantum gates make it possible to implement algorithms that fully utilize the capabilities of quantum physics by utilizing the special characteristics of qubits and their interactions.

CHAPTER 3

ALGORITHMS IN QUANTUM

Compared to classical algorithms, quantum algorithms solve problems more quickly by utilizing the ideas of quantum physics. This chapter examines some of the most important quantum algorithms, such as Shor's and Grover's algorithms, and talks about how they affect many other domains, especially database search and cryptography. We also look at new quantum algorithms and how they might affect quantum computing in the future.

3.1 The Algorithm of Shore

One of the most well-known quantum algorithms is Shor's algorithm, which was created in 1994 by mathematician Peter Shor. It is intended to factor huge integers exponentially more quickly than the most well-known classical techniques, which has important cryptographic ramifications.

The time required by classical techniques, such as the general number field sieve, increases exponentially with the number of digits in the number being factored. This results in an exponential speedup in factoring. In particular, classical approaches are inefficient for very big numbers due to their sub-exponential temporal complexity. Shor's approach, on the other hand, runs in polynomial time, i.e. $(O((\log N)^2 (\log \log N) (\log N)))$, where (N) is the number that needs to be subtracted. This indicates that Shor's approach can locate factors far faster than traditional techniques for large integers.

The algorithm can be divided into a number of crucial steps:

- **Finding the Quantum Period:** The quantum Fourier transform, which determines the period of a function related to modular exponentiation, is the fundamental component of Shor's algorithm. The method is able to take use of quantum superposition and interference because of this stage.
- **Classical Post-Processing:** The factoring process is finished by computing the factors from the period

information using traditional techniques after the period has been established.

The impact on cryptography is as follows:

Since RSA encryption depends on the difficulty of factoring large integers as its security foundation, the ability to factor huge numbers effectively poses a serious threat to traditional cryptographic systems. A sufficiently potent quantum computer might crack RSA encryption, jeopardizing global data security and secure communication infrastructures. In order to create new encryption techniques that are resistant to quantum attacks, this understanding has sparked research into quantum-resistant cryptographic algorithms.

3.2 The Algorithm of Governor

Lov Grover developed the Grover's technique in 1996, which provides a quadratic speedup for searching unsorted databases. This algorithm solves a basic computer science problem: how to efficiently search through a list of (N) things.

- **Quadratic Speedup:** In classical computing, it takes (O(N)) time in the worst scenario to search through an unsorted database of (N) elements because each item needs to be examined separately. This complexity is reduced to $(O(\sqrt{N}))$ by Grover's technique, which makes it possible for the program to find the target item much more quickly, particularly in huge datasets.

Grover's algorithm can be summed up in the following steps:

- **Initialization**: Every state that could exist is set up in superposition.
- **Query to the Oracle:** Among the options, a "oracle" function selects the right answer and marks it without disclosing its value.
- Through a sequence of iterations that use both the oracle and a diffusion operator, the algorithm amplifies the probability amplitude of the correct solution.
- **Measurement:** The qubits are measured following a sufficient number of iterations, and the right answer is most likely recovered.

Grover's Algorithm Applications:

Grover's algorithm has a wide range of possible uses in numerous fields:

1. **Cryptography:** It can be used to cut the required key length in half, hence attacking symmetric key cryptography systems. Using Grover's approach, for example, a 256-bit key would offer security on par with a 128-bit key against a quantum opponent.

2. **Optimization Issues:** A lot of optimization issues can be expressed as search issues. Applying Grover's approach can improve the effectiveness of locating the best answers in intricate datasets.

3. **Machine Learning:** Grover's algorithm can enhance some machine learning tasks, especially those that require looking for patterns or anomalies in big datasets.

3.3 The Potential of Other Quantum Algorithms

Numerous new quantum algorithms, each with distinct capabilities and consequences for the future of quantum computing, are being created in addition to Shor's and

Grover's algorithms.

The QFT, or quantum fourier transform, is:
A key element of many quantum algorithms, such as Shor's algorithm, is the quantum Fourier transform, which is a quantum variant of the classical discrete Fourier transform. Effective frequency analysis made possible by QFT makes it possible for quantum computers to complete some jobs far more quickly than their classical equivalents.

QFT Applications:
1. **Signal Processing:** QFT can improve the effectiveness of filtering and reconstructing signals in applications like image processing.
2. **Quantum Simulations:** QFT is essential for modeling quantum systems and examining how they behave over time, supporting studies in materials science and chemistry.

Emerging Quantum Algorithms: The study of quantum algorithms is a fast-moving topic, and several new algorithms are showing promise for a range of uses:
- The Variational Quantum Eigensolver (VQE) is a

hybrid approach that finds a Hamiltonian's lowest eigenvalue by combining classical and quantum computing. Because it enables researchers to examine molecule structures and processes, this approach is very pertinent to quantum chemistry.

- **The algorithm known as the Quantum Approximate Optimization Algorithm (QAOA):** By using quantum superposition and entanglement to investigate possible solutions more quickly than traditional algorithms, this approach is intended to solve combinatorial optimization issues.
- **The HHL Algorithm:** This algorithm, named for its developers Harrow, Hassidim, and Lloyd, solves linear equation problems tenfold more quickly than traditional techniques. It is particularly pertinent to machine learning and optimization applications.

Implications for Quantum Computing's Future:

Quantum algorithms will change computational capabilities in a variety of industries as they continue to advance. The ability to resolve intricate issues that are now beyond the capabilities of traditional computers creates new opportunities in domains including machine learning, materials science, cryptography, and optimization.

Quantum algorithms are a revolutionary development in computational methodology that make it possible to solve issues that were previously thought to be impossible or extremely difficult. The future of technology will be shaped by the continued study and advancement of these algorithms, underscoring the importance of quantum computing in solving challenging problems in a variety of fields. As time goes on, the cooperation of quantum theory and real-world application will propel computer breakthroughs, encouraging creativity and novel approaches to problem-solving.

CHAPTER 4

SECURITY AND QUANTUM CRYPTOGRAPHY

The field of cryptography faces previously unheard-of opportunities and challenges as quantum computing technology develops. This chapter explores post-quantum cryptography protocols, the novel concept of Quantum Key Distribution (QKD), and the dangers that quantum computers pose to traditional encryption techniques. Comprehending these ideas is essential as we get ready for a future in which secure communication may be completely transformed by quantum technologies.

4.1 Classical Cryptography and Quantum Dangers

Significant flaws in classical cryptography systems are brought about by the development of quantum computers, especially when those systems rely on mathematical issues that are currently challenging for classical computers to solve.

Breaking Classical Encryption Protocols: The strength of certain mathematical problems specifically, factoring large integers and solving discrete logarithm problems, respectively is the foundation of classical encryption protocols like RSA (Rivest-Shamir-Adleman) and ECC (Elliptic Curve Cryptography).

These issues can be resolved by quantum computers in polynomial time by employing methods such as Shor's algorithm:

- **RSA Encryption:** The difficulty of factoring a product of two big prime integers is the key to RSA's security. Shor's algorithm essentially makes RSA susceptible by enabling a quantum computer to factor these numbers tenfold quicker than traditional techniques.
- The difficulty of the elliptic curve discrete logarithm problem is the foundation of elliptic curve cryptography (ECC). Shor's technique can also target ECC, making it possible for a quantum computer to efficiently calculate the private key from the public key.

Timeline Projections: There is disagreement among experts over the anticipated availability of quantum computers that may crack various cryptographic methods. Although estimates differ, some predict that within the next ten to thirty years, usable quantum computers may become available:

- **Current Developments:** Only a small number of qubits are now in use, and quantum computers are still in the experimental stage. The quantity of qubits and their coherence durations should, however, greatly increase as technology develops.
- **Preparation for Quantum Threats:** In anticipation of the unavoidable advent of sufficiently powerful quantum computers, organizations and governments are encouraged to begin the transition to quantum-resistant cryptographic protocols.

In conclusion, there is a genuine and immediate quantum danger to traditional cryptography, which calls for immediate information security conversations and actions.

4.2 Post-Quantum Cryptography

The area of cryptography is rapidly developing quantum-resistant protocols, or post-quantum cryptography, in response to the dangers posed by quantum computers.

The purpose of post-quantum cryptography techniques is to protect against the possible capabilities of quantum computers. These protocols, in contrast to traditional approaches, are based on mathematical issues that are challenging for both classical and quantum computers:

1. **Lattice-Based Cryptography:** Lattice problems are thought to be immune to quantum assaults and entail determining the shortest vector in a high-dimensional lattice. Lattice cryptography-based protocols include Learning With Errors (LWE) and NTRU.
2. **Hash-Based Signatures:** These depend on hash functions' security, which is often impervious to quantum attacks. Hash-based signature techniques include Merkle signatures.
3. **Code-Based Cryptography:** This type of

cryptography depends on how difficult it is to decode random linear codes. One well-known and thoroughly researched code-based cryptosystem is McEliece.

The difficulty of solving systems of multivariate polynomial equations over finite fields is the foundation of the Multivariate Quadratic Equations method.

In terms of assessing and standardizing post-quantum cryptography algorithms, the National Institute of Standards and Technology (NIST) is at the forefront of these efforts. In order to gather, examine, and suggest quantum-resistant algorithms, the organization has implemented a multi-phase process:

- The Post-Quantum Cryptography Standardization project, which NIST started in 2016, underwent multiple evaluation rounds. NIST seeks to set future standards by evaluating candidate algorithms according to their security, effectiveness, and implementation features.
- As of 2022, NIST has announced the first set of algorithms to be standardized, which includes hash-based and lattice-based schemes. These

initiatives are crucial for directing governments and businesses toward the use of safe cryptography techniques in the quantum era.

Post-quantum cryptography is a proactive strategy for protecting data from the possible risks posed by quantum computing, making sure that security protocols develop in tandem with new developments in technology.

4.3 Quantum Key Distribution (QKD)

A novel technique called Quantum Key Distribution (QKD) builds a secure communication channel by applying the ideas of quantum physics. QKD ensures the security of the transmitted key, in contrast to traditional key distribution techniques that may be susceptible to interception.

How QKD Operates: To guarantee secure communication, QKD depends on the characteristics of quantum states. The BB84 protocol, created in 1984 by Charles Bennett and Gilles Brassard, is the most well-known QKD protocol. Several crucial steps are involved in the process:

- The process of preparation and transmission involves Alice, the sender, preparing quantum bits (qubits) in different states and sending them to Bob, the recipient. There are four different polarization states for each qubit: horizontal, vertical, diagonal, and anti-diagonal.
- To measure each qubit that is received, Bob selects a basis (collection of states) at random. Any attempt by Eve, the eavesdropper, to intercept the qubits will disrupt their states due to the nature of quantum mechanics, making her presence visible.
- **Key Generation:** Alice and Bob compare their measurement bases for the sent qubits following the transmission. In order to create a shared secret key that may be utilized for encryption, they discard any qubits whose bases did not match.
- **Eavesdropping Detection:** Eve will unavoidably create errors if she attempts to measure the qubits. Alice and Bob can determine whether their key is secure by comparing a subset of their key bits to look for any differences.

Practical Applications of QKD: QKD is being actively studied and applied in a number of fields:

1. **Banking and Finance:** To ensure that sensitive data is shielded from possible breaches, financial institutions are investigating QKD for safe communications and transactions.
2. **Government and Military Applications**: QKD is regarded as essential for maintaining national security and private communications through military and government channels, protecting classified data.
3. **Telecommunications:** To give customers better data transfer security, telecommunications organizations are looking into integrating QKD into their current infrastructure.

Current Challenges of QKD: Despite its potential, there are still a number of obstacles to overcome before QKD may be used in practice.

- **Distance Limitations:** Because quantum signals attenuate over long distances, QKD systems usually have a restricted range. There are ongoing efforts to

use quantum repeaters to increase this range.
- The technology needed for QKD can be expensive and complicated to implement, which could prevent its broad use, especially in smaller businesses.
- **Integration with Classical Systems:** There are extra technological problems when integrating QKD with current classical cryptography systems, necessitating hybrid solutions for efficient data security.

Quantum key distribution is a ground-breaking development in secure communication that provides strong defense against intrusions and attacks. With further research and development, QKD could revolutionize secure data transmission in a society growing more interconnected by the day.

There are both serious risks and potential at the nexus of encryption and quantum computing. It is critical that people, businesses, and governments modify their security plans as we traverse this complicated terrain in order to protect private data from quantum threats and investigate the cutting-edge possibilities that quantum cryptography presents. In order to ensure secure communication in the quantum age, the development of QKD and the continuous growth of cryptographic protocols will be crucial.

CHAPTER 5

DRUG DISCOVERY USING QUANTUM COMPUTING

Drug development and quantum computing are coming together at a critical juncture in the pharmaceutical business that could have revolutionary effects. Conventional drug development techniques frequently need a lot of time and resources, but the special powers of quantum computers hold the potential to improve accuracy and efficiency. This chapter examines how the simulation of molecular structures, expedited drug design and testing, and expanded applications of quantum computing into genomics and biotechnology can transform the drug discovery process.

5.1 Molecular Structure Simulation

Accurately modeling intricate chemical interactions and structures is one of the biggest obstacles in drug discovery. Because quantum systems are so complicated, classical computers are inherently limited in this area.

Restrictions of Classical Computers: When replicating quantum phenomena, classical computers' use of bits that represent either a 0 or a 1 causes exponential scaling problems. Larger systems are computationally impossible to mimic due to the drastically increased number of possible configurations and interactions for complex molecules. Important restrictions consist of:

- **Exponential Complexity:** As a molecule's atom count rises, the computational demands on classical simulations climb exponentially. For example, there may be more combinations in a molecule with only 100 atoms than there are atoms in the visible universe.
- **Approximation Methods:** In order to overcome these constraints, traditional approaches frequently use approximations, which may produce less precise findings. Essential quantum effects like electron correlation and wavefunction overlap may be missed by methods like Monte Carlo methods or molecular dynamics simulations.

Quantum Computing for molecule Modeling: Using the ideas of quantum mechanics, quantum computers may more precisely and effectively model molecule interactions:

- The ability of qubits to represent several states at once enables quantum computers to investigate numerous chemical configurations in parallel. Studying quantum systems, where superposition and entanglement are essential, benefits greatly from these capabilities.

- **Accurate Energy Calculations:** Molecular energies and structures can be optimized with previously unheard-of accuracy using quantum algorithms like the Variational Quantum Eigensolver (VQE) and Quantum Approximate Optimization Algorithm (QAOA). These algorithms effectively investigate energy landscapes by utilizing quantum parallelism.

In order to comprehend reaction mechanisms and drug-target interactions, researchers must be able to watch how molecules interact over time. Quantum computing can offer precise insights into molecular dynamics.

Potential medication candidates could be found more

quickly and with less reliance on trial-and-error techniques because of quantum computers' capacity to correctly mimic molecular structures.

5.2 Quickening Drug Development and Evaluation

Drug discovery is a notoriously drawn-out process that frequently costs billions of dollars and takes more than ten years. These durations could be greatly shortened by quantum computing, which would simplify the medication development and testing procedures.

Increasing Drug Discovery Timelines: Quantum computing can improve the drug discovery process at many levels, from preliminary screening to clinical trials:

- **Quick Compound Screening:** Compared to traditional techniques, quantum algorithms are more effective at analyzing large libraries of chemical compounds. For instance, researchers can find good candidates for more research significantly more quickly by employing Grover's technique, which can search unsorted databases quadratically faster.
- Early in the discovery process, better informed

design choices can be made by using quantum simulations to forecast how effectively a therapeutic molecule would attach to its target protein. The amount of chemicals that require laboratory synthesis and testing is decreased by this predictive capability.

- **Reducing Experimental Cycles:** Researchers may virtually refine molecular designs through quantum-enhanced simulations, which eliminates the need for physical synthesis and testing. Development times can be greatly shortened by this decrease in trial cycles.

Several groups are now investigating the use of quantum computers in drug discovery. Here are some examples of early-stage quantum applications in pharmaceutical research:

- IBM has collaborated with pharmaceutical businesses to create quantum algorithms for drug development, which are known as IBM's Quantum Applications. In order to find possible inhibitors for illnesses like Alzheimer's, one noteworthy effort involves modeling how chemicals bind to proteins.

- The chemical structures of possible drug candidates have been optimized using D-Wave's quantum annealers, which have yielded insights that may help find more potent drugs.
- In order to increase the precision of molecular property predictions and improve the drug discovery process by gaining a deeper understanding of chemical behavior, Google is researching quantum algorithms.

These early-stage uses highlight the growing interest in and promise of quantum computing to revolutionize the pharmaceutical sector and open the door to quicker and more effective drug discovery procedures.

5.3 Biotechnology and Genomics: Quantum Applications

Beyond drug development, quantum computing has an impact on the larger domains of biotechnology and genetics, where its capabilities can help with some of the most important problems.

The following are some ways that quantum algorithms can improve genomic research:

- Applications of Quantum Computing in Gene Sequencing and Bioinformatics
- **Efficient Sequence Alignment:** One of the core tasks in genomics is aligning DNA and protein sequences. Understanding genetic variants requires the speed and accuracy of sequence matching, which can be enhanced by quantum computers using methods like the Quantum Fourier Transform.
- **Data Analysis in Genomics**: Researchers can extract insights from intricate biological datasets far more quickly than traditional computers can handle thanks to quantum computing techniques, which can effectively analyze the massive volumes of data produced by genomic investigations.

Long-Term Implications for Personalized Medicine: By facilitating more accurate and customized treatment plans, quantum computing holds the potential to completely transform personalized medicine:

- **Targeted Drug Development:** Researchers can create targeted treatments based on each patient's unique genetic profile by using quantum simulations to comprehend patient-specific molecular interactions. This move to individualized medicine is expected to decrease side effects and increase treatment effectiveness.

- **Integrating Genomic Data with Quantum Algorithms:** Combining genomic data and quantum computing can help find new drug targets and provide groundbreaking treatments tailored to individual genetic compositions.

In conclusion, quantum computing applications in genetics and biotechnology offer insights that could result in major improvements in healthcare and individualized treatment plans, expanding the possibilities for creative solutions beyond medication development.

In conclusion, because of its capacity to model intricate chemical structures, speed up design procedures, and extend its reach into biotechnology and genomics, quantum computing has the potential to revolutionize drug discovery and associated domains. The pharmaceutical sector stands to gain much from advances in research and the development of quantum technologies, which could change the way we approach personalized medicine and medication development in the years to come. The revolutionary potential of quantum computing is emphasized in this chapter, highlighting the necessity of ongoing research and investment in this fascinating area of science and technology.

CHAPTER 6

OPTIMIZATION PROBLEMS WITH QUANTUM COMPUTING

From supply chain management and logistics to finance, optimization issues are prevalent in many industries. When confronted with the size and complexity of contemporary problems, especially those involving complicated constraints or the analysis of large datasets, traditional optimization techniques sometimes fail. A new paradigm for addressing these issues is provided by quantum computing, which makes use of the ideas of quantum physics to solve problems more quickly. This chapter explores how quantum computing can be used to optimize supply chains, logistics, and financial portfolios. It also explores how quantum machine learning and optimization can work together.

6.1 Financial Optimization

Advanced optimization techniques have the potential to

revolutionize the financial sector. Financial companies can increase profitability and make better decisions by using quantum computing to improve trading methods, risk analysis, and portfolio management.

Optimizing Portfolio Management: Choosing the ideal asset mix to minimize risk and attain a desired return is the goal of portfolio management. Conventional approaches frequently use heuristic algorithms and linear programming, which can become computationally complex as the number of assets rises. This procedure can be greatly enhanced by quantum computing via:

- **Quantum Algorithms for Portfolio Optimization:** Quantum algorithms, like the Variational Quantum Eigensolver (VQE) and the Quantum Approximate Optimization Algorithm (QAOA), may effectively explore multiple asset combinations and assess their possible risks and rewards at the same time. Financial managers can use quantum superposition and entanglement to better optimize portfolios thanks to these capabilities.
- **Real-time Market Adaptation:** Quantum computers are able to dynamically modify portfolios

by analyzing market data in real-time. Investors can swiftly determine the best asset allocations and react nimbly to market swings by analyzing big datasets of market situations.

Early Use Cases in Financial Modeling and Risk Analysis: To improve their modeling and risk assessments, a number of financial organizations are starting to test quantum computing:

- The investment bank JP Morgan has been investigating the application of quantum algorithms for pricing complicated derivatives and portfolio optimization. Their goal is to use quantum computing to decrease the processing time involved in financial modeling and enhance risk assessments.
- **Barclays with D-Wave:** Barclays has shown the potential for quicker and more precise financial modeling by working with D-Wave to explore how quantum computing might be used to optimize options pricing and enhance risk management techniques.
- **IBM Quantum with Financial Institutions:** IBM Quantum has obtained access to quantum hardware

and software for financial algorithm prototyping and has collaborated with a number of financial institutions to investigate the potential uses of quantum computing in risk analysis and portfolio optimization.

Quantum computing has the potential to revolutionize the financial industry by enhancing risk analysis and portfolio management, underscoring its significance as a key instrument for upcoming financial innovation.

6.2 Logistics and Supply Chain Optimization

Routing, scheduling, and resource allocation are some of the most difficult optimization problems that the logistics and supply chain industries must deal with. These issues can be creatively solved by quantum computing, which helps companies cut expenses and increase productivity.

Difficult Routing and Scheduling Issues: There are many variables in logistics, including delivery locations, time limits, and vehicle capacities. When these variables are dynamic or interconnected, traditional optimization methods could have trouble identifying the best answers. These complications can be addressed by quantum computing by:

- **Quantum Algorithms for Route Optimization:** The Vehicle Routing Problem (VRP) and the Traveling Salesman Problem (TSP) may both be solved efficiently by quantum algorithms. Quantum annealing approaches, for example, enable the simultaneous investigation of several paths, producing optimal or nearly ideal solutions much more quickly than conventional methods.

- **Dynamic Scheduling:** Schedules can be dynamically modified by quantum computing in response to real-time data, including delivery requirements and traffic conditions. This flexibility improves operational effectiveness by guaranteeing appropriate resource allocation and maintenance of service levels.

Practical Uses in Manufacturing and Transportation: Businesses are starting to apply quantum computing solutions to enhance their supply chain and logistics processes:

- **Volkswagen's Traffic Optimization:** Volkswagen has developed algorithms to optimize traffic flow in urban areas in partnership with quantum computing businesses. They hope to lessen gridlock and enhance car routing by analyzing traffic patterns with quantum computing.
- **DHL and Quantum Algorithms**: In order to cut costs and enhance service quality, DHL is investigating the use of quantum computing in its logistics operations. The company is concentrating on streamlining supply chain procedures including inventory control and delivery scheduling.
- **Honeywell's Quantum Solutions for Manufacturing:** Honeywell has created quantum algorithms that speed up production scheduling and supply chain logistics, increasing productivity and cutting waste.

Quantum computing's incorporation into supply chain management and logistics can result in more effective operations, lowering expenses and improving service quality.

6.3 Optimization via Quantum Machine Learning

There is great potential for resolving challenging optimization issues at the nexus of machine learning and quantum computing. Innovative answers to a range of problems can be obtained using hybrid approaches that combine the advantages of quantum computing with traditional machine learning techniques.

Hybrid Approaches to Problem-Solving: By fusing conventional machine learning with quantum algorithms, improved models that can solve optimization problems that are now unsolvable can be created:

- **Quantum Neural Networks:** These models improve on conventional neural networks by utilizing the concepts of quantum computing. Quantum neural networks may handle information more effectively by utilizing quantum properties like superposition. This enables faster convergence and increased accuracy in optimization tasks.
- **Quantum Support Vector Machines (QSVM):** QSVMs improve support vector machine

classification performance, especially in high-dimensional areas, by applying quantum principles. In optimization scenarios with a large and complex search space, this skill might be quite important.

Uses in Complex Problem-Solving: Quantum machine learning is already being investigated in a number of domains to forecast outcomes and optimize procedures:

- In the realm of finance and trading, hybrid quantum-classical algorithms are being developed to improve trading techniques through faster and more efficient analysis of market data to find winning transactions.
- **Healthcare:** By examining intricate biological data, quantum machine learning can improve the identification of biomarkers or drug candidates in genomics and drug discovery.
- In order to optimize energy grid management, enable more effective energy distribution, and lower operating costs, quantum machine learning techniques are being researched in the Energy Sector.

Combining machine learning and quantum computing offers a potent toolkit for solving optimization issues, improving the capacity to extract useful information from intricate models and datasets.

Quantum computing offers a revolutionary chance to optimize a number of sectors, such as machine learning, banking, and logistics. Organizations may solve complicated optimization issues more quickly by utilizing quantum algorithms, which improves operational performance, lowers costs, and facilitates better decision-making. Applications of quantum computing in optimization are expected to grow as research and technology progress, spurring innovation and changing sectors in the process. This chapter emphasizes how crucial quantum computing is to resolving some of the most difficult optimization issues of our day.

CHAPTER 7

Physical Implementation and Quantum Hardware

The successful application of quantum bits, or qubits, the basic building blocks of quantum information, is essential to the realization of quantum computing. Qubits have been created using a variety of physical systems, each with special benefits and difficulties. This chapter examines the most popular qubit implementation technologies, such as trapped ions and superconducting circuits, in addition to additional cutting-edge qubit technologies. We will also discuss the major difficulties in developing quantum hardware, such as maintaining coherence, reducing noise, and pursuing error correction.

7.1 Qubits that are superconducting

Because of their comparatively high coherence times and capacity to combine numerous qubits into scalable systems, superconducting qubits have become one of the most promising applications of quantum computing.

Superconducting Qubit Mechanism:
Josephson junctions, which are thin insulating barriers between two superconductors, are the foundation of superconducting qubits. These junctions have the ability to represent both 0 and 1 at the same time because they can behave quantumly and create superposition states. Superconducting qubits come in the following primary varieties:

1. **Transmons:** a kind of superconducting qubit that improves coherence times by reducing sensitivity to charge noise. Transmons are very popular due to their scalability and ease of manufacture.
2. **Flux Qubits:** These qubits can be adjusted for a variety of uses and represent qubit states using magnetic flux; nevertheless, they are more susceptible to noise in the environment than transmons.

Companies Leading the Way in Superconducting Qubits: A number of IT firms are leading the way in the development of superconducting qubit technology:

- **IBM:** Leading the way in superconducting qubit research and development is IBM Quantum, the company's quantum computing program. Their dedication to developing quantum computing is demonstrated by their Quantum Experience platform, which enables customers to use their quantum processors for experimentation and algorithm testing.
- **Google:** With their Sycamore processor, which achieved quantum supremacy in 2019, Google has made notable progress in superconducting qubits. This significant achievement showed that their quantum computer could outperform the most potent classical supercomputer in the world at a particular task.
- This company, Rigetti Computing, specializes in creating superconducting qubit devices and provides cloud-based quantum computing services that let users test their technology.

Superconducting qubits continue to be a fundamental component of quantum computing, and their scalability and performance are being driven by ongoing developments.

7.2 Other Qubit Technologies and Trapped Ions

Although the field is dominated by superconducting qubits, trapped ions are another well-known physical qubit implementation. Ions contained by electromagnetic fields are used in trapped ion systems, which enable accurate measurement and manipulation.

The advantages and disadvantages of trapped ions with relation to superconducting qubits are as follows:

The following are advantages:
1. **High Fidelity:** By isolating ions from ambient noise, trapped-ion systems can operate with remarkably high fidelity and lengthy coherence durations.
2. Because trapped ions can be stacked in a variety of ways, scalable designs that effortlessly integrate numerous qubits are made possible. This is known as "natural scalability."

The following are challenges:

1. **Complexity of Control:** It is difficult to scale processes to large numbers of qubits because the precise control of trapped ions necessitates intricate laser systems and exact timing.
2. **Speed Limitations:** In some applications, the performance of trapped ions may be limited due to their slower operations compared to superconducting qubits.

Other qubit technologies are being researched in addition to superconducting qubits and trapped ions. These include:

1. **Topological Qubits:** To accomplish fault-tolerant quantum computation, these qubits make use of strange particles known as anyons. Although actual implementations are still in their infancy, theoretical research is still ongoing and holds promise for reliable qubit systems.
2. **Qubits that are photonic:** Through different states of light (such as polarization), photons can be utilized as qubits. Although measurement and integration issues still exist, photonic systems have the advantage of being intrinsically less prone to decoherence.

3. **Quantum Dots:** It is possible to design semiconductor structures at the nanoscale to act as qubits. Electronic manipulation is possible with quantum dots, but scalability and coherence times are issues.

In order to overcome the difficulties presented by each, research into trapped ions and other qubit technologies is still ongoing. The goals are to increase operating efficiency, fidelity, and scalability.

7.3 Quantum Hardware Challenges

The development of practical quantum computers is hampered by a number of important issues, despite the encouraging developments in quantum hardware. In order to create scalable and dependable quantum computing systems, these issues must be resolved.

Preserving Coherence and Minimizing Noise: Decoherence, in which qubit states lose their quantum characteristics as a result of interactions with their surroundings, is a well-known vulnerability of quantum systems. Long coherence times are necessary to carry out intricate quantum calculations:

1. **Environmental Noise:** A variety of noise sources, including thermal fluctuations and electromagnetic interference, can impact qubits. Scientists are looking at ways to protect qubits from these disruptions.

2. **Materials Science Innovations:** Purer and more stable qubit systems are the goal of materials science developments. The performance and coherence times of superconducting qubits, for example, can be improved by advancements in superconducting materials.

Creating Error-Corrected Quantum Systems: Since even minor mistakes can result in major computational failures, error correction is essential for realistic quantum computing. Error-corrected quantum systems are being constructed using a variety of methods:

1. It is possible to detect and rectify faults without physically measuring the qubits thanks to quantum error correction codes. Well-known codes, such the surface code, necessitate a large overhead because

they use several physical qubits to express a single logical qubit.

2. **Fault-Tolerant designs:** Scientists are creating designs that are error-proof by design, enabling quantum algorithms to run dependably. The purpose of these systems is to control and reduce errors that occur during quantum operations.

The road towards useful quantum hardware will become more obvious as scientists continue to tackle these issues, opening the door for reliable and expandable quantum computing solutions.

The foundation of quantum computing technology is represented by quantum hardware and its physical implementations. Every technology, from trapped ions to superconducting qubits and other cutting-edge technologies, has advantages and disadvantages of its own. The field's advancement depends on the continuous research efforts to create efficient error correction systems, lower noise, and improve coherence. The potential of quantum computing to transform industries and resolve challenging issues becomes more apparent as we move toward more dependable and scalable quantum systems. The complex terrain of quantum hardware development and the exciting developments influencing the direction of quantum computing are depicted in this chapter.

CHAPTER 8

PROGRAMMING LANGUAGES AND QUANTUM SOFTWARE

The creation of software that can efficiently utilize the capabilities of quantum hardware is becoming more and more important as the area of quantum computing advances. Programming languages, frameworks, and tools created especially for the special characteristics of quantum systems are all included in the category of quantum software. This chapter explores the best practices for developing quantum algorithms, the burgeoning quantum software ecosystem, and the top quantum programming frameworks. Developers can more effectively traverse the intricate terrain of quantum computing and aid in its progress by being aware of these elements.

8.1 Frameworks for Quantum Programming

The infrastructure required to write and run quantum algorithms is provided by quantum programming frameworks. There are now a number of noteworthy platforms that serve various purposes and capacities in the field of quantum computing.

Overview of Leading Quantum Programming Platforms:

Qiskit: One of the most well-known open-source quantum computing frameworks, Qiskit was created by IBM. With IBM's quantum hardware, it allows users to design, model, and run quantum circuits. Qiskit is made up of multiple parts:

- The core layer for creating and executing quantum programs is **Terra**, which provides tools for circuit design, optimization, and translation.
- **Aer:** A powerful quantum circuit simulator that lets programmers test algorithms without requiring actual quantum hardware.
- **Ignis:** A collection of tools for characterisation and error mitigation that aid users in comprehending and enhancing the functionality of their quantum systems.

An open-source framework called Cirq was created by

Google with the express purpose of building, modeling, and running quantum circuits on NISQ (Noisy Intermediate-Scale Quantum) devices. Important characteristics include:

- The flexibility of Cirq lies in its low-level approach to quantum circuit design, which allows for precise control over quantum gates and measurements.
- **Integration with TensorFlow:** Cirq easily combines with TensorFlow, enabling developers to efficiently blend quantum and classical operations.

Braket: Provided by Amazon Web Services (AWS), Braket is a fully managed quantum computing solution that works with D-Wave and Rigetti, among other quantum hardware backends. Among the noteworthy features are:

- Braket's high-level interface makes it simple for users to create and execute quantum algorithms on a variety of quantum devices.
- **Hybrid Workflows:** Braket enables users to optimize their algorithms by utilizing both quantum computing and classical resources through hybrid quantum-classical workflows.

How They Differ from Classical Programming Languages: Because quantum programming frameworks reflect the following essential laws of quantum mechanics, they differ greatly from classical programming languages like Python or Java:

- **Superposition and Entanglement:** Quantum programming permits qubits to reside in superposition, allowing for more sophisticated data representations than classical programming, which depends on binary states (0s and 1s).
- A distinct approach to logic and control flow is required because quantum gates change qubit states probabilistically, in contrast to conventional functions that carry out deterministic operations.
- **Probabilistic Outputs:** Because quantum programs produce probabilistic results, their creators must create algorithms that take statistical behavior and measurement uncertainty into consideration, which is not possible with classical programming.

These distinctions call for a different approach to problem-solving, and programmers must modify their talents to fully take advantage of quantum computing's special potential.

8.2 Ecosystem of Quantum Software

The ecosystem for quantum software is developing quickly and includes a range of platforms, tools, and libraries intended to make it easier to create and run quantum algorithms on quantum hardware.

The Emerging Ecosystem of Software Tools: A number of libraries and software tools have been created to improve the functionality of frameworks for quantum programming:

Before implementing quantum algorithms on actual quantum hardware, it is essential to test them in a simulated setting using quantum simulators. Developers can diagnose and refine their algorithms by using simulators, which replicate the behavior of qubits and quantum gates. Well-known quantum simulators consist of:

- **ProjectQ:** An open-source framework that allows users to effectively mimic quantum circuits by integrating with various backends, such as IBM's

Qiskit.
- A simulator dedicated to simulating quantum algorithms, specifically variational quantum algorithms, is called QuISP.

Quantum Algorithm Libraries: To make the process of creating quantum applications easier, specialized libraries have been created to offer pre-built quantum algorithms and components. Among the examples are:

- **PennyLane:** A quantum machine learning library that interfaces with current machine learning frameworks to make it simple for developers to create and train quantum models.
- **OpenFermion:** This library, created by Google, offers tools for simulating electronic structures and focuses on quantum computing applications in quantum chemistry and materials research.

Quantum Simulators and Their Significance in Present Development: Quantum simulators are essential to the advancement of quantum software development since they fulfill a number of crucial roles:

- **Algorithm Testing and Validation:** Before

implementing their quantum algorithms on actual hardware, developers can test them in a controlled setting to make sure they work properly.

- **Resource Estimation:** By assisting in the estimation of the resources (such as qubit count and gate operations) needed to execute particular quantum algorithms, simulators allow for more effective hardware use.
- **Understanding Quantum Behavior**: Developers can better understand how quantum systems behave and create efficient algorithms by modeling different quantum events.

These tools' incorporation into the ecosystem of quantum software encourages cooperation and creativity, opening the door for fresh uses and developments in quantum computing.

8.3 The Development of Quantum Algorithms

It takes a special method and comprehension of quantum mechanics as well as the unique properties of quantum hardware to create efficient quantum algorithms. The development of quantum computing is being aided by the emergence of best practices and strategies.

Optimal Methods and Techniques for Creating Quantum Algorithms:

The following best practices should be taken into account by developers while creating quantum algorithms:

- The first step is to have a classical understanding. Numerous quantum algorithms are extensions of their conventional counterparts. Designing new quantum solutions and spotting possible quantum speedups can be made easier with an understanding of classical algorithms.
- Make Use of Well-Known Quantum Algorithms: Use well-known quantum algorithms as building blocks for more intricate applications, such as Grover's and Shor's algorithms. Combining and modifying these algorithms can result in creative fixes for particular issues.
- **Pay Attention to Problem Decomposition:** Divide difficult issues into smaller, more manageable parts that can be solved with quantum methods. The design process can be made simpler and specific

optimizations made possible by this decomposition.

The following tools are available to help developers make the switch from classical to quantum programming: Tools and Frameworks that Help Developers Transition to Quantum Programming

1. **Educational Resources:** Developers can learn quantum concepts and programming paradigms by using online tutorials, classes, and documentation offered by companies including Google, Microsoft, and IBM.
2. **Community and Collaboration Platforms:** Qiskit Community and Quantum Computing Stack Exchange are two examples of online forums and collaborative platforms that promote knowledge sharing and offer assistance to developers who need help or comments on their projects.
3. **Integrated Development Environments (IDEs):** Developers can more easily experiment with quantum computing thanks to specialized IDEs for quantum programming, including Qiskit Notebooks and Amazon Braket Studio, which provide integrated environments for creating, testing, and

visualizing quantum algorithms.

Developers will need to modify their procedures and adopt new tools and approaches that support efficient quantum algorithm creation as quantum computing develops further.

The development of quantum programming languages and software is a crucial part of the revolution in quantum computing. Leading frameworks for quantum programming, such as Qiskit, Cirq, and Braket, offer special features that let developers use quantum computing for a range of purposes. With its simulators and specialized libraries, the new quantum software ecosystem promotes creativity and improves the development process. Developers can efficiently contribute to the quick development of quantum technology by using the tools that are available and adhering to best practices for creating quantum algorithms. In order to fully realize the potential of quantum computing, this chapter emphasizes the importance of quantum software and exhorts developers to learn more about and work in this fascinating area.

CHAPTER 9

QUANTUM COMPUTING APPLICATIONS IN INDUSTRY

By using its special ability to analyze enormous volumes of data, solve challenging issues, and simulate complex systems that are much beyond the scope of classical computing, quantum computing has the potential to completely transform a wide range of sectors. This chapter examines the revolutionary possibilities of quantum computing in the fields of artificial intelligence (AI), healthcare, and energy, stressing its present uses, constraints, and potential future developments.

9.1 Medical Applications of Quantum Computing

Significant advancements in a number of healthcare domains, including genomics, diagnostics, and medical research, could result from quantum computing. Large datasets and intricate biological systems might be analyzed at previously unheard-of rates, perhaps resulting in breakthroughs that improve patient outcomes and expedite medical procedures.

Potential Advancements in Medical Research, Genomics, and Diagnostics: Quantum computing has the potential to solve a number of issues facing the medical field, including:

- **Drug Discovery:** Conventional methods of finding new drugs might take years and use a lot of computing power. Because quantum computers can accurately model chemical interactions, they may be able to find promising medication candidates much more quickly. For example, scientists may simulate ligand-receptor interactions and protein folding, offering insights that were previously impossible to obtain with traditional computation.
- **Genomic Sequencing:** Deciphering enormous volumes of information is a necessary part of analyzing genomic data. Faster detection of genetic markers linked to diseases is made possible by the optimization of genome sequencing using quantum algorithms. Approaches to personalized medicine, in which a patient's therapies are customized based on

their genetic composition, may result from this.

- **Diagnostics:** By examining intricate datasets from genetic testing, medical imaging, and electronic health records, quantum machine learning approaches might improve diagnostic procedures. For instance, by spotting patterns that classical computers might overlook, quantum algorithms may increase the precision of tumor identification in imaging scans.

Despite the enormous potential of quantum computing in healthcare, there are still a number of obstacles to overcome:

- **Technological Readiness**: Present-day quantum computers are still in their infancy, with low error rates and qubit counts that make them unsuitable for real-world use. A significant obstacle to widespread adoption is still achieving the required gate integrity and coherence times.
- **Privacy and Security of Data:** Since sensitive patient data is used more and more in healthcare, protecting its security and privacy is crucial. These issues might be resolved by quantum cryptography,

but incorporating it into current systems calls for careful thought.

- **Regulatory Obstacles:** Regulatory frameworks that might not be prepared to manage the subtleties of quantum computing will need to be navigated in order to utilize quantum technology in the healthcare industry.

Notwithstanding these drawbacks, further study and funding in quantum technologies have the potential to revolutionize healthcare by improving patient outcomes and streamlining procedures.

9.2 Energy Quantum Computing

Managing grid stability, switching to renewable energy sources, and streamlining supply chains are just a few of the many issues facing the energy industry. These problems can be creatively solved by quantum computing, which will result in more effective energy systems and aid in the battle against climate change.

The following are some ways that quantum computing can

improve the energy sector:

Applications in Optimizing Energy Grids and Improving Renewable Energy Technologies

The management of energy grids entails the resolution of intricate optimization issues pertaining to load forecasting, resource allocation, and supply and demand balancing. Large datasets from sensors and smart meters can be analyzed by quantum algorithms to enhance grid operations in real-time, reducing energy loss and enhancing dependability.

- **Renewable Energy Integration:** Because renewable energy sources like wind and solar are intermittent, integrating them presents difficulties. In order to ensure that renewable energy is used effectively, quantum computing can assist in the development of predictive models that optimize energy distribution and storage. Quantum algorithms, for instance, can enhance grid resilience by improving predicting models for wind and solar energy generation.
- **Material Science for Energy Applications:** New materials for energy conversion and storage, such

fuel cells and batteries, can be found more quickly thanks to quantum computing. Researchers can find materials with greater efficiency and longer lifespans by quantum-level modeling material characteristics.

The Role of Quantum in Addressing Climate Change: As the globe struggles with climate change, quantum computing offers a chance to create novel solutions:

- In order to assist combat climate change, quantum simulations can be used to develop more effective techniques for capturing and storing carbon dioxide. Quantum technologies have the potential to revolutionize carbon capture technology by better understanding molecular interactions and streamlining chemical reactions.
- The development of sustainable energy solutions, such as improved solar cells and more effective catalysts for the creation of hydrogen, can be aided by quantum computing. These developments could facilitate the shift to a low-carbon economy and lessen dependency on fossil fuels.

Quantum computing has the potential to greatly help the

energy industry by facilitating the creation of novel, sustainable solutions as well as by optimizing current systems.

9.3 Artificial Intelligence and Quantum Computing

Numerous industries are being drastically changed by artificial intelligence, and quantum computing may hasten AI research and applications. Developers can improve data analysis, tackle challenging optimization problems, and improve machine learning algorithms by utilizing the special qualities of quantum systems.

How Quantum Computing Could Hasten AI Research: There are numerous ways in which quantum computing can have a big impact on AI.

- **Speeding Up Training:** It usually takes a lot of time and processing power to train machine learning models. Training procedures can be accelerated by quantum algorithms, including the Quantum Approximate Optimization Algorithm (QAOA), which allows for quicker convergence and better model performance.

- **Enhanced Data Analysis**: Compared to traditional methods, quantum machine learning is more effective at analyzing huge datasets, uncovering insights and patterns that would otherwise go undetected. High-dimensional data can be processed using quantum algorithms, which enhances the accuracy of AI applications' predictions and classifications.

Quantum Algorithms and Machine Learning Synergies: Research on the nexus of quantum computing and machine learning is ongoing, and the results show promising synergies:

- **Hybrid Quantum-Classical Models:** AI applications may perform better when quantum and classical methods are combined. For example, hybrid models can leverage quantum systems for complicated pattern identification while using conventional pre-processing to address data noise.
- **Quantum Neural Networks:** By utilizing quantum entanglement and superposition, researchers are investigating the creation of quantum neural networks that may perform better than classical

neural networks. These networks might be useful for applications like natural language processing and picture recognition.

- **Reinforcement Learning**: Quantum algorithms may improve reinforcement learning, which would enable agents to more efficiently explore and take advantage of their surroundings. Quantum computing has the potential to improve learning processes by enabling quicker assessments of actions and results.

Quantum computing could open up new avenues for AI research, speeding up developments and paving the way for the creation of more complex and powerful AI systems.

Quantum computing has the potential to revolutionize a number of sectors, such as artificial intelligence, healthcare, and energy. There are countless opportunities for innovation and advancement in these fields as long as scientists and companies keep investigating and developing quantum technologies. Even if there are still issues with data privacy, technological maturity, and regulations, the continued interest in and investment in quantum computing point to a bright future. Businesses may solve difficult problems, improve operational effectiveness, and help create a more inventive and sustainable world by utilizing the special powers of quantum systems.

CHAPTER 10

QUANTUM COMPUTING'S FUTURE

Quantum computing has a bright and complicated future ahead of it, with the potential to revolutionize many industries by resolving issues that traditional computers are currently unable to handle. This chapter addresses the long-term vision of quantum supremacy and its ramifications, looks at the recent advancements and near-term milestones in quantum technology, and talks about the ethical and societal issues that arise from the widespread use of quantum computing.

10.1 Recent Developments and Upcoming Milestones

Recent years have seen tremendous progress in quantum computing, from abstract ideas to real developments in hardware, software, and applications. The capabilities of quantum systems will be significantly enhanced by a number of significant milestones that are expected to occur soon as research progresses.

Key Developments in Quantum Computing in the Past Few Years: A number of noteworthy advancements have occurred in the field of quantum computing, including:

- **Increased Qubit Counts:** As they go from small-scale systems to more complicated designs, companies such as IBM, Google, and Rigetti have dramatically increased the number of qubits in their quantum processors. Google, for example, has developed a 72-qubit processor known as "Bristlecone," and IBM has stated that it would have a 1,121-qubit machine dubbed "Condor" by 2023.

- **Improved Coherence Times:** Longer coherence times, which enable qubits to sustain their quantum state for prolonged periods of time, are the result of improvements in qubit design and materials. This is essential for carrying out intricate algorithms and doing accurate computations.

- **Error Correction Techniques:** The fidelity of quantum operations has increased with the invention of quantum error correction codes, such as surface codes and color codes. These methods are crucial for

reducing the effects of noise and guaranteeing the precision of quantum calculations.
- **Quantum Software Development:** The development of quantum algorithms and applications has been made easier by the appearance of quantum programming languages and frameworks like Qiskit, Cirq, and Pennylane. These platforms promote innovation in the field by allowing developers and academics to test and prototype quantum algorithms.

Key Breakthroughs Anticipated in the Next 5-10 Years:
A number of noteworthy advances in quantum computing are anticipated in the near future.
- In order to show that quantum computers are superior to their classical counterparts at certain activities, researchers are working to demonstrate quantum advantage in real-world applications. This milestone is expected to be reached in fields like material science, drug development, and optimization.
- One of the top priorities is the creation of scalable quantum systems that can be used in commercial settings. In order to run more complex algorithms,

this entails not just adding more qubits but also making sure that they are connected and controlled effectively.

- **Integration with Classical Systems:** It is probable that quantum computing and classical systems will be integrated in the future, resulting in hybrid designs that capitalize on their respective advantages. A more seamless transition to quantum-enhanced applications across multiple industries will be made possible by this combination.

- **Standardization and Interoperability:** As the ecosystem surrounding quantum computing develops, attempts to standardize protocols and guarantee interoperability across various quantum systems will gain momentum. Collaboration and the creation of universal quantum applications will be made easier as a result.

These developments paved the way for a quickly changing quantum environment that might open up new possibilities and revolutionize entire sectors.

10.2 Future Prospects: Quantum Domination and Beyond

The point at which a quantum computer can do a calculation that is impossible for any classical computer, regardless of its capabilities, is known as quantum supremacy. Although reaching quantum supremacy is an important milestone, the route ahead is far more expansive.

The Idea of Quantum Supremacy and Its Consequences
In 2019, Google asserted quantum supremacy when its 53-qubit quantum processor, Sycamore, finished a particular task in about 200 seconds, a time that would have taken the most potent classical supercomputer in the world 10,000 years to finish. This accomplishment has multiple ramifications:

- Validation of Quantum Principles: The principles of quantum physics and the potential of quantum computing as a game-changing technology are validated by proving quantum supremacy.
- **Catalyst for Research and Investment:** The declaration of quantum supremacy sparked interest from the public and corporate sectors, leading to a

surge in research and investment in quantum technologies. It is anticipated that this momentum will spur other developments and creativity in the area.

- **Practical Applications:** Although quantum supremacy in and of itself does not lead to practical applications, it opens the door to investigating real-world issues where quantum computing can offer significant benefits, as in sophisticated simulations, optimization, and cryptography.

What Takes Place After Reaching Quantum Supremacy: Reaching quantum supremacy is not the end of the quest. Rather, the emphasis will move to:

- **Creating Useful Quantum Applications:** Following their demonstration of superiority, scientists will focus on creating quantum algorithms that solve real-world issues in a variety of fields, including logistics, healthcare, and finance.
- **Improving Coherence and Error Rates:** The pursuit of improving coherence times and mistake correction will remain crucial. The development of fault-tolerant quantum computing will be necessary

to carry out increasingly intricate and dependable algorithms.

- **Broadening Accessibility:** Innovation and a variety of quantum technology applications will be greatly aided by making quantum computing available to a wider audience via cloud platforms and educational programs.

Beyond the idea of dominance, the long-term goal of quantum computing is to create a strong ecosystem of applications that use quantum principles to address some of the most difficult problems facing the planet.

10.3 Implications for Society and Ethics

Quantum computing technologies will unavoidably have important societal and ethical ramifications as they develop and proliferate. It will be crucial to comprehend and resolve these issues in order to guarantee the responsible development and application of quantum technology.

The Social Impact of Widespread Deployment of Quantum Computing: The society will be profoundly impacted by the widespread deployment of quantum computing.

- **Transforming Industries:** As quantum technologies allow for unprecedented levels of efficiency, precision, and personalization, sectors like healthcare, banking, and logistics will experience significant changes. Along with possible disruptions in traditional professions, this shift may result in the creation of jobs in domains relevant to quantum technology.

- **Economic disadvantages:** As quantum computing develops, differences in access to knowledge and technology could surface, which could exacerbate already-existing disadvantages. To guarantee that everyone in society benefits equally, efforts must be made to democratize access to quantum resources and knowledge.

- **National Security Concerns:** Existing cryptography systems are challenged by quantum computing's capabilities, which raises national security issues. A move to quantum-resistant cryptography is required to protect sensitive data because increasingly potent quantum computers have the ability to crack popular encryption techniques.

Ethical Issues in Quantum Research and Its Possible Disruptions to Different Industries: The quick development of quantum computing brings up moral issues that need to be resolved.

- **Responsible Research Practices:** To guarantee accountability, transparency, and integrity in the creation of quantum technology, ethical standards for quantum research ought to be set. Developing a framework that encourages responsible innovation will require cooperation between ethicists, researchers, and legislators.
- **Impact on Privacy:** There are serious privacy issues because quantum computing has the ability to crack current encryption. Strong security measures and a reassessment of privacy rules will be necessary to guarantee that personal data is secured in a quantum future.
- **Balancing Innovation with Caution:** The quest for quantum computing innovations should be balanced with prudence, especially when it comes to applications that could affect individual rights and

society standards. Discussions regarding the moral ramifications of their work and the possible effects on society as a whole are essential for stakeholders.

Quantum computing's future involves more than just technical development; it also involves larger societal issues that call for careful consideration and discussion amongst multiple stakeholders.

Quantum computing has a bright future ahead of it, with recent developments paving the way for game-changing breakthroughs in a variety of sectors. It is critical to take into account the societal and ethical ramifications of these advancements as scientists strive toward immediate milestones and investigate the long-term vision of quantum supremacy. By encouraging a responsible and inclusive approach to quantum technology, we may utilize its potential to solve difficult problems and improve society.

ABOUT THE AUTHOR

Technology specialist and author Emma Royce Smartley specializes in the newest developments in AI, coding tools, and software development. His goal as a writer is to help developers and tech fans remain ahead of the curve by simplifying difficult tech ideas. His publications provide insightful analyses of how new technologies are changing the landscape of productivity and software development. Emma's love of innovation propels him to investigate and elucidate the technologies that will shape the landscape of the future.

www.ingramcontent.com/pod-product-compliance
Lightning Source LLC
Chambersburg PA
CBHW050323230526
45471CB00005B/2316